Everyday Mathematics

The University of Chicago School Mathematics Project

Student Math Journal
Volume 1

Grade **1**

Mc Graw Hill **Wright Group**

The University of Chicago School Mathematics Project (UCSMP)

Max Bell, Director, UCSMP Elementary Materials Component; Director, *Everyday Mathematics* First Edition
James McBride, Director, *Everyday Mathematics* Second Edition
Andy Isaacs, Director, *Everyday Mathematics* Third Edition
Amy Dillard, Associate Director, *Everyday Mathematics* Third Edition

Authors

Max Bell	Robert Hartfield	Kathleen Pitvorec
Jean Bell	Andy Isaacs	Peter Saecker
John Bretzlauf	James McBride	
Amy Dillard	Rachel Malpass McCall*	

*Third Edition only

Technical Art
Diana Barrie

Teachers in Residence
Jeanine O'Nan Brownell
Andrea Cocke
Brooke A. North

Editorial Assistant
Rossita Fernando

Contributors

Cynthia Annorh, Robert Balfanz, Judith Busse, Mary Ellen Dairyko, Lynn Evans, James Flanders, Dorothy Freedman, Nancy Guile Goodsell, Pam Guastafeste, Nancy Hanvey, Murray Hozinsky, Deborah Arron Leslie, Sue Lindsley, Mariana Mardrus, Carol Montag, Elizabeth Moore, Kate Morrison, William D. Pattison, Joan Pederson, Brenda Penix, June Ploen, Herb Price, Dannette Riehle, Ellen Ryan, Marie Schilling, Susan Sherrill, Patricia Smith, Kimberli Sorg, Robert Strang, Jaronda Strong, Kevin Sweeney, Sally Vongsathorn, Esther Weiss, Francine Williams, Michael Wilson, Izaak Wirzup

Photo Credits

©C Squared Studios/Getty Images, p.34; ©Ralph A. Clevenger/CORBIS, cover, *center;* Getty Images, cover, *bottom left;* ©Tom and Dee Ann McCarthy/CORBIS, cover, *right;* ©Stockdisc/Getty Images, p.31.

www.WrightGroup.com

Wright Group

Send all inquiries to:
Wright Group/McGraw-Hill
P.O. Box 812960
Chicago, IL 60681

ISBN 0-07-604535-8

6 7 8 9 CPC 12 11 10 09 08 07

Contents

UNIT 3 Visual Patterns, Number Patterns, and Counting

Measurement and Basic Facts

UNIT 5 Place Value, Number Stories, and Basic Facts

Activity Sheets

Koala
19 lb

Cheetah
120 lb

Penguin
75 lb

LESSON 1·4 Number Writing: 1

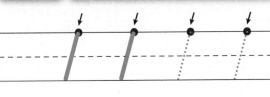

	1 + 0	⊡
1		
/	2 − 1	
uno		one

Draw a picture of 1 thing.

LESSON 1·4 Number Writing: 2

	1 + 1	⊡
2		
//	3 − 1	
dos		two

Draw a picture of 2 things.

Date

Number Writing: 3

Draw a picture of 3 things.

3

$2 + 1$

/// $4 - 1$

tres three

Number Writing: 4

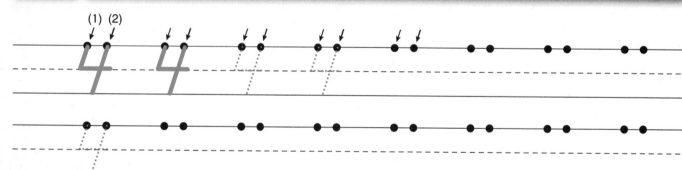

(1) (2)

Draw a picture of 4 things.

4

$3 + 1$

//// $5 - 1$

cuatro four

LESSON 1·8   Dice-Roll and Tally

Roll a die. Use tally marks to record the results on this chart.

	Tallies	Total

Calendar

Month

Sunday	Monday	Tuesday	Wednesday	Thursday	Friday	Saturday

LESSON 1·9 Number Writing: 5

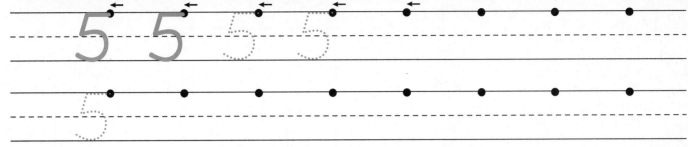

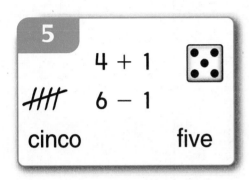

Draw a picture of 5 things.

LESSON 1·9 Number Writing: 6

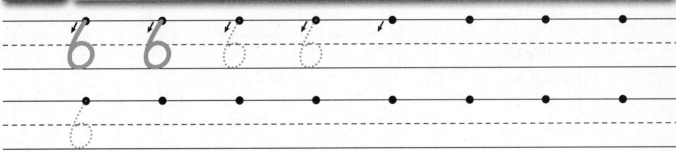

6

5 + 1

||||| | 7 − 1

seis   six

Draw a picture of 6 things.

LESSON 1·12 A Thermometer

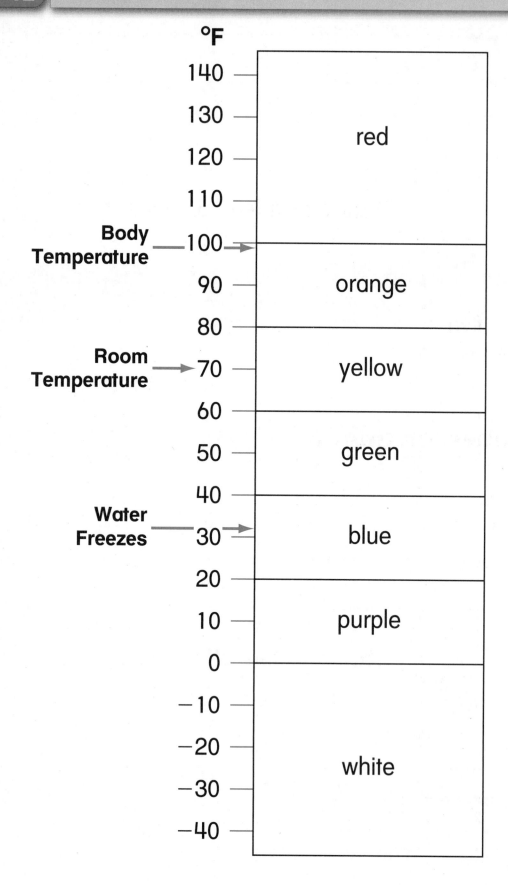

Rolling for 50

Materials

◆ a die

◆ a marker for each player

◆ a gameboard

Players 2

Skill Count by 1s

Object of the Game

To be the first player to reach 50

Directions

Take turns.

1. Put your marker on 0.

2. Roll the die. Look in the table to see how many spaces to move.

3. The first player to reach 50 wins.

Roll	Spaces
1	3 up
2	2 back
3	5 up
4	6 back
5	8 up
6	10 up

| | | | | | | | | | 0 |

1	2	3	4	5	6	7	8	9	10
11	12	13	14	15	16	17	18	19	20
21	22	23	24	25	26	27	28	29	30
31	32	33	34	35	36	37	38	39	40
41	42	43	44	45	46	47	48	49	50

LESSON 2·2 — Information about Me

My first name is _____.

My second name is _____.

My last name is _____.

I am _____ years old.
Put candles on your cake.

My area code and home telephone number are

(_____ _____ _____) _____ _____ _____ – _____ _____ _____ _____

    (area code)   (telephone number)

Important Phone Numbers

Emergency number: _____ _____ _____
School number:

(_____ _____ _____) _____ _____ _____ – _____ _____ _____ _____

Local library number:

(_____ _____ _____) _____ _____ _____ – _____ _____ _____ _____

LESSON 2·2   Number Writing: 7

7 7 7 7

7

6 + 1

~~HHHt~~ //

siete

8 − 1

seven

Draw a picture of 7 things.

LESSON 2·2   Number Writing: 8

8 8 8 8 8

8

7 + 1

~~HHHt~~ ///

ocho

9 − 1

eight

Draw a picture of 8 things.

nine **9**

Math Boxes

---

1. Count up by 1s.

7,   8,   9,

————, ————, ————,

————, ————, ————,

————, ————

---

2. Count up by 5s.

0,   5,   10,

————, ————, ————,

————, ————, ————

---

3. Write the number that comes before.

——— 19

——— 23

——— 31

——— 36

---

4. Write the number.

————   ————

 **LESSON 2·4**   **Number Writing: 9**

9	
8 + 1	
~~HHI~~ ////	10 − 1
nueve	nine

Draw a picture of 9 things.

 LESSON 2·4   **Number Writing: 0**

0	
0 + 0	☐
1 − 1	
cero	zero

LESSON 2·4 Math Boxes

1. How many tally marks?

~~HHT~~ ~~HHT~~ //

Choose the best answer.

- ⬭ 3
- ⬭ 12
- ⬭ 15
- ⬭ 10

2. Draw the shape you are more likely to grab from the bag.

_____

3. Make a sum of 10 pennies.

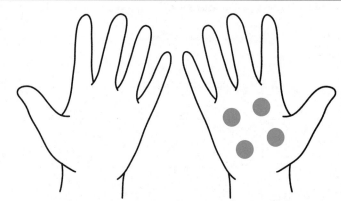

4. Complete the number line.

8   9   ___ ___ ___ ___ ___ ___

Math Boxes

1. Count back by 1s.

18,   17,   16,

——— , ——— , ——— ,

——— , ——— , ——— ,

——— , ———

2. Count up by 5s.

10,   15,   20,

——— , ——— , ——— ,

——— , ———

3. What number comes before 10?

Choose the best answer.

 1

⬭ 0

⬭ 9

⬭ 11

4. Write the number.

———   ———

LESSON 2·6 **Telling Time**

1. Record the time.

_____ o'clock

_____ o'clock

_____ o'clock

_____ o'clock

2. Draw the hour hand.

2 o'clock

6 o'clock

LESSON 2·6 Math Boxes

1. How many tally marks?

卌 卌 卌 ////

_____ tally marks

2. Draw the shape you are more likely to grab from the bag.

_____

3. Make a sum of 10 pennies.

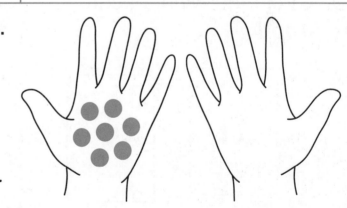

4. Complete the number line.

22   23   ___   ___   ___   ___   ___

Date _____

1. Record the time.

_____ o'clock

2. Use your number grid.

Start at 12.

Count up 5.

You end at _____.

3. Circle the winning card in *Top-It*.

11   9

4. What day of the week is today?

_____

What day of the month?

_____

What day of school?

_____

LESSON 2·8   **Math Boxes**

---

1. Use your number grid.

Start at 11.

Count back 6.

You end at _____.

2. How much money?

_____ ¢

---

3. What comes next?

Choose the best answer.

---

4. Count up by 2s.

0,   2,   4, _____, _____, _____,

_____, _____, _____, _____, _____

LESSON 2·9   **Exploring Pennies and Nickels**

Write the total amount. Then show the amount using fewer coins.
Write ⓟ for penny and Ⓝ for nickel.

(*Hint:* Exchange pennies for nickels.)

1. ⓟ ⓟ ⓟ ⓟ ⓟ ⓟ ⓟ   _____¢

Show this amount using fewer coins.

2. ⓟ ⓟ ⓟ ⓟ ⓟ ⓟ ⓟ ⓟ ⓟ   _____¢

Show this amount using fewer coins.

3. ⓟ ⓟ ⓟ ⓟ ⓟ ⓟ ⓟ ⓟ ⓟ ⓟ ⓟ ⓟ   _____¢

Show this amount using fewer coins.

Try This

4. Ⓝ ⓟ ⓟ ⓟ ⓟ ⓟ ⓟ   _____¢

Show this amount using fewer coins.

**LESSON
2·9** **Math Boxes**

1. Record the time.

_____ o'clock

2. Use your number grid.

Start at 15.

Count up 9.

You end at _____.

Choose the best answer.

◯ 6

◯ 15

◯ 24

◯ 23

3. Circle the winning card in *Top-It*.

| 19 | 9 |

4. What day of the week is today?

_____

What day of the month?

_____

What day of school?

_____

LESSON 2·10 Counting Pennies and Nickels

Write the total amount.

1.

_____ ¢

2.

_____ ¢

3.

_____ ¢

Try This

4.

```
P  N  P    P  P
P        P  N    P
   P  P  P  P
```

Write the total amount. _____ ¢

Show this amount using fewer coins.

Math Boxes

1. Use your number grid.

Start at 18.

Count back 8.

You end at _____.

2. How much money?

Ⓟ Ⓟ Ⓟ Ⓟ Ⓟ Ⓟ Ⓟ Ⓟ

Choose the best answer.

⬭ 10¢

⬭ 8¢

⬭ 16¢

⬭ 9¢

3. Draw what comes next.

_____ _____

_____ _____

4. Count up by 2s.

2,   4,   6,   _____, _____,

_____, _____, _____, _____, _____

Math Boxes

1. How much money?

N N P

_____ ¢

2. Draw the hour hand.

2 o'clock

3.

The Pets We Own	
Pet	Tallies
Cat	ⵘⵘ ⵘⵘ
Dog	ⵘⵘ ///
Other	ⵘⵘ /

How many cats? _____ cats

How many dogs?

_____ dogs

4. Count up by 10s.

20,   30,   40,

_____, _____, _____,

_____, _____, _____,

_____, _____

LESSON 2·12 **Math Boxes**

1. Use your number grid.

Start at 23.

Count up 8.

You end at _____.

2. How much money?

Ⓝ Ⓝ Ⓟ Ⓟ Ⓟ

_____ ¢

3. Draw what comes next.

   _____ _____

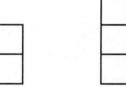   _____ _____

4. Count up by 2s.

6,   8,   10,   _____,

_____, _____, _____, _____

School Store Mini-Poster 1

crayon
9¢

scissors
10¢

ball
25¢

gum
1¢

pencil
18¢

candy
5¢

eraser
7¢

Counting Coins

1. Tell how much money.

_____ ¢

_____ ¢

How much money in all? _____ ¢

2. Buy 2 items from the School Store. Draw them below.

3. Under each item you drew, show how much it costs.
   Use Ⓟ for pennies and Ⓝ for nickels.

4. Circle the item that costs more.
   How much more does it cost? _____ ¢

Try This

5. Draw 2 items that cost a total of 14¢.

Math Boxes

1. Use Ⓟ and Ⓝ to show the cost.

28¢

_____

2. It is _____ o'clock.

Choose the best answer.

- ⬭ 12
- ⬭ 8
- ⬭ 6
- ⬭ 7

3. Fill in the table.

The Pets We Own		
Pet	Tallies	Total
Cat	⃥⃥⃥⃥ ///	
Dog	⃥⃥⃥⃥ ⃥⃥⃥⃥ //	
Other	⃥⃥⃥⃥ ////	

4. Count back by 10s.

90,   80,   70,   _____, _____, _____,

_____, _____, _____

Math Boxes

1. Write the number.

_____   _____

2. Count up by 10s.

30,   40,   _____,

_____, _____, _____,

_____, _____

3. Use your template. Make a pattern with ◯s and △s.

4. Complete the number line.

_____ 10   11   _____ _____ _____ _____

LESSON 3·1 **Patterns**

1. Draw the next 2 shapes.
   Use your Pattern-Block Template.

    ___ ___

    ___ ___

     ___ ___

2. Make up your own pattern.
   Then ask your partner to draw the next 2 shapes.

Try This

3. Draw the next 3 shapes.

LESSON 3·1 — Math Boxes

1. Circle the winning card in *Top-It*.

| 18 | | 17 |

2. Draw the hour hand.

6 o'clock

3. Record the total amount.

Ⓟ Ⓟ Ⓟ Ⓟ Ⓟ Ⓟ

_____ ¢

Use Ⓟ and Ⓝ to show this amount with fewer coins.

4. What is the temperature?

Fill in the circle next to the best answer.

Ⓐ about 60°F

Ⓑ about 40°F

Ⓒ about 70°F

Ⓓ about 50°F

°F
60—
50—
40—
30—
20—
10—

Date

How many ☐s? Label **odd** or **even**.

Example:

___4___

even

1.

_____

_____

2.

_____

_____

3.

_____

_____

How many ☆s? Label **odd** or **even**.

4.

_____   _____

Try This

5.

_____

Math Boxes

1. Use Ⓟ and Ⓝ to show the cost.

2. What shape comes next?

Fill in the circle next to the best answer.

    ____

Ⓐ   Ⓑ ☐

Ⓒ △   Ⓓ ◯

3. Make sums of 10 pennies.

Left Hand	Right Hand
9	1
4	
	5

4. Complete this part of the number grid.

1	2			5
		13	14	
21				25
	32		34	

The 2s Pattern

									0
1	2	3	4	5	6	7	8	9	10
11	12	13	14	15	16	17	18	19	20
21	22	23	24	25	26	27	28	29	30
31	32	33	34	35	36	37	38	39	40
41	42	43	44	45	46	47	48	49	50
51	52	53	54	55	56	57	58	59	60
61	62	63	64	65	66	67	68	69	70
71	72	73	74	75	76	77	78	79	80
81	82	83	84	85	86	87	88	89	90
91	92	93	94	95	96	97	98	99	100
101	102	103	104	105	106	107	108	109	110

Shade the 2s pattern on the above grid.

Fill in the missing numbers below.

_0_ , _2_ , _4_ , ____ , _8_ , ____ ,

____ , _14_ , ____ , ____ , _20_ , ____ ,

____ , ____ , _28_ , ____ , ____ , _34_

Date _____

1. Circle the winning card in *Top-It.*

| 22 | 18 |

2. Draw the hour hand.

4 o'clock

3. Record the total amount.

_____ ¢

Use Ⓝ to show this amount with fewer coins.

4. Color the thermometer to show about 40°F.

°F
50—
40—
30—
20—
10—

MURDOCK LEARNING RESOURCE CENTER

Math Boxes

1. Use Ⓟ and Ⓝ to show the cost.

2. Draw what comes next.

_____ _____

_____ _____

3. Make sums of 10 pennies.

Left Hand	Right Hand
3	7
4	
	8

4. Complete this part of the number grid.

		8		10
	17			
	27	28	29	
36				

LESSON 3·5 **Number-Line Skip Counting**

1. Show counts by 2s.

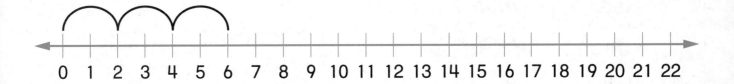

2. Show counts by 5s.

3. Show counts by 10s.

Try This

4. Show counts by 3s.

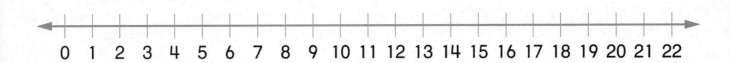

Math Boxes

1.

Favorite Colors	
Red	~~HHH~~ ~~HHH~~
Green	~~HHH~~ ~~HHH~~ //
Yellow	~~HHH~~ ////
Blue	~~HHH~~ ~~HHH~~ ~~HHH~~

Which color is most popular?

Fill in the circle next to the best answer.

Ⓐ red   Ⓒ yellow

Ⓑ green   Ⓓ blue

2. Count by 5s. Circle your counts on the number grid.

1	2	3	4	5	6	7	8	9	10
11	12	13	14	15	16	17	18	19	20
21	22	23	24	25	26	27	28	29	30
31	32	33	34	35	36	37	38	39	40
41	42	43	44	45	46	47	48	49	50

3. How many ☐s? _____

Odd or even? _____

4. Complete the number line.

____ 28 29 ____ ____ ____ ____ ____

Adding and Subtracting on the Number Line

0 1 2 3 4 5 6 7 8 9 10 11 12 13 14 15 16 17 18 19 20 21 22 23 24 25

1. Start at 6. Count up 2 hops. Where do you end up?

$6 + 2 =$ _____

2. Start at 4. Count up 9 hops. Where do you end up?

$4 + 9 =$ _____

3. Start at 15. Count back 7 hops. Where do you end up?

$15 - 7 =$ _____

4. Start at 18. Count back 8 hops. Where do you end up?

$18 - 8 =$ _____

Try This

5. $5 + 8 =$ _____

6. $11 - 8 =$ _____

7. $3 + 13 =$ _____

LESSON 3·6 **Math Boxes**

1. Complete the table.

Before	Number	After
11	12	13
	8	
	15	
	19	

2. How many days are in a week?

Fill in the circle next to the best answer.

Ⓐ 5

Ⓑ 7

Ⓒ 10

Ⓓ 30

3. Count up by 2s.

12,   14,   16,

_____, _____, _____,

_____, _____, _____,

_____, _____

4. Circle the longer one.

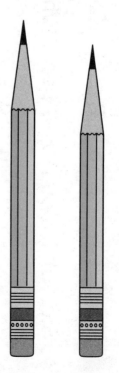

LESSON 3·7 **Telling Time**

1. Record the time.

half-past _____ o'clock   half-past _____ o'clock

_____ o'clock   _____ o'clock

2. Draw the hour hand and the minute hand.

half-past 1 o'clock   7 o'clock

LESSON 3·7 Math Boxes

1.

Weather	
Sunny	ⵌ ⵌ ⵌ ///
Cloudy	ⵌ ⵌ ////
Rainy	ⵌ ⵌ
Snowy	ⵌ ////

How many sunny days?

_____ sunny days

Were there more rainy days or snowy days?

more _____ days

2. Count up by 5s.

__15__, __20__, __25__, _____, _____, _____,

_____, _____, _____

3. How many ☐s? _____

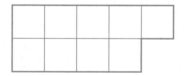

Odd or even? _____

4. Make sums of 10 pennies.

Left Hand	Right Hand
2	8
5	
	9

Date _____

Frames and Arrows

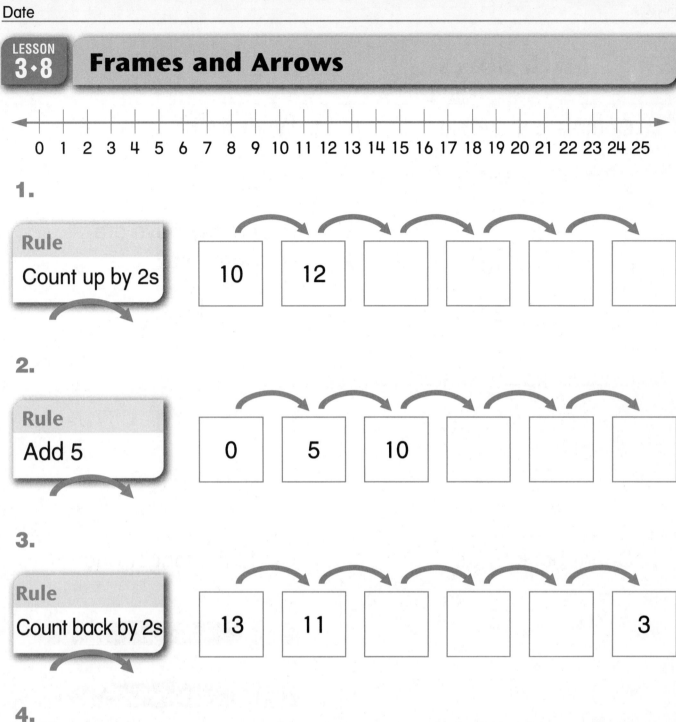

0 1 2 3 4 5 6 7 8 9 10 11 12 13 14 15 16 17 18 19 20 21 22 23 24 25

1.

Rule
Count up by 2s

| 10 | 12 | | | | |

2.

Rule
Add 5

| 0 | 5 | 10 | | | |

3.

Rule
Count back by 2s

| 13 | 11 | | | | 3 |

4.

Rule
Subtract 3

| 15 | | | | | |

Date

Math Boxes

1. Complete the table.

Before	Number	After
7	8	9
	10	
	20	
	29	

2. What month is it?

_____

How many days are
in this month?

_____ days

3. Count back by 2s.

20,   18,   16,

_____, _____, _____,

_____, _____, _____,

_____, _____

4. Circle the longer one.

Date _____

1. Fill in the frames.

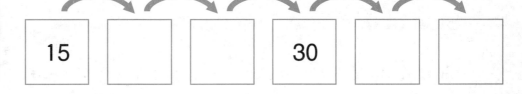

Rule						
Count by 5s Add 5		15			30		

2. Fill in the rule.

Rule	20	18	16	14	12	10

3. Fill in the rule and the frames.

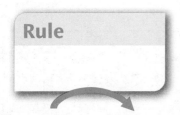

Rule	7	10	13			

4. Fill in the rule and the frames.

 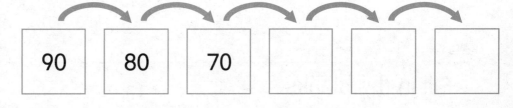

Rule	90	80	70			

5. Make up your own.

 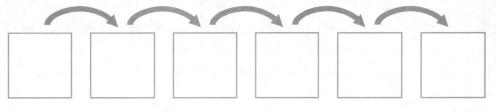

Rule						

LESSON 3·9 Math Boxes

1. Use your number line.

Start at 0.

Count up 6.

You end at _____.

$0 + 6 =$ _____

2. Record the time.

half-past _____ o'clock

3.

Rule
Count by 5s

 15

4. Fill in the blanks.

_10_ , _20_ , _30_ , _____ , _50_ ,

_____ , _____ , _80_ , _____ , _____

Date

1. Odd or even?

_____   _____

_____   _____

2. Make sums of 10 pennies.

Left Hand	Right Hand
7	3
4	
	8

3. Use a calculator.
Count up by 3s.

   0,   3,   6,

_____, _____, _____,

_____, _____, _____,

_____, _____

4. What is the number model?

Fill in the circle next to the best answer.

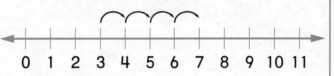

0 1 2 3 4 5 6 7 8 9 10 11

Ⓐ 3 + 5 = 7

Ⓑ 4 + 3 = 7

Ⓒ 3 − 4 = 7

Ⓓ 3 + 4 = 7

Coin Exchange

Show each amount using fewer coins.
Write Ⓟ for penny, Ⓝ for nickel, and Ⓓ for dime.

1.

2.

3.

4.

Math Boxes

---

1. Use your number line.
Start at 3.
Count up 5.
You end at _____.

Fill in the circle next to the best answer.

Ⓐ 3

Ⓑ 5

Ⓒ 7

Ⓓ 8

2. What time is it?

half-past _____ o'clock

---

3. Fill in the rule and the missing numbers.

Rule

| 5 | 10 | | | |

---

4. Count up by 10s.

__20__ , __30__ , __40__ , _____ , _____ ,

_____ , _____ , _____ , _____ , _____

LESSON 3·12 How Much Money?

How much money?
Use your coins.

P	N	D
1¢	5¢	10¢
$0.01	$0.05	$0.10
penny	nickel	dime
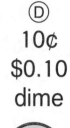		

Example:

_____35_____ ¢

How much money? Use your coins.

1.   _____ ¢

2.   _____ ¢

3.   _____ ¢

Try This

4. Ⓓ Ⓓ Ⓝ Ⓝ Ⓝ Ⓟ Ⓟ _____ ¢

LESSON 3·12 Math Boxes

1. Use your number line.
Start at 8.
Count back 5.

You end at _____.

8 − 5 = _____

2. Draw the hour hand and the minute hand.

half-past 12 o'clock

3. How much money?

Ⓓ Ⓓ Ⓓ Ⓝ Ⓟ Ⓟ

Fill in the circle next to the best answer.

Ⓐ 6¢

Ⓑ 22¢

Ⓒ 30¢

Ⓓ 37¢

4. Write the number model.

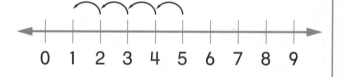

_____

LESSON 3·13 **Math Boxes**

1. Odd or even?

_____   _____

2. Make sums of 10 pennies.

Left Hand	Right Hand
8	2
6	
	3

3. Use a calculator.
Count up by 7s.

0,   7,   14,

_____ , _____ , _____ ,

_____ , _____ , _____ ,

_____ , _____

4. Write the number model.

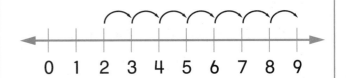

0 1 2 3 4 5 6 7 8 9

_____

Date _____

Domino Parts and Totals

Write 3 numbers for each domino.

Example:

Total
6

Part	Part
4	2

1.

Total

Part	Part

2.

Total

Part	Part

3.

Total

Part	Part

4.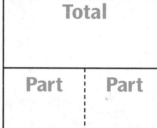

Total

Part	Part

5.

Total

Part	Part

6. Draw dots in the domino. Write 3 numbers in the diagram.

Total

Part	Part

7. Find the missing part. Draw dots in the domino.

Total
8

Part	Part
3	

Date _____

1. Use your number line.
Start at 6.
Count back 4.

You end at _____.

6 − 4 = _____

2. Draw the hour hand and the minute hand.

half-past 2 o'clock

3. Write the total amount.

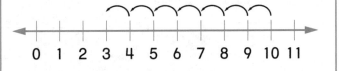

_____ ¢

4. Write the number model.

0 1 2 3 4 5 6 7 8 9 10 11

_____

52   fifty-two

Date

1. Color the thermometer to show about 75°F.

°F
80—
70—
60—
50—
40—

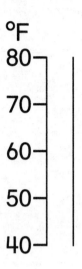

2. Circle the longer one.

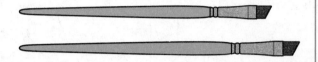

3. Complete this part of the number grid.

1	2			5
11	12			15
		23		25
			34	

4. Count up by 10s.

___50___ , ___60___ ,

___70___ , _____ ,

_____ , _____ ,

_____ , _____ ,

_____ , _____

LESSON 4·1 **Reading Thermometers**

°F

120 —
110 —
Body
Temperature 100 —
90 —
80 —
Room 70 —
Temperature
60 —
50 —
40 —
Water
Freezes 30 —
20 —
10 —
0 —
−10 —
−20 —
−30 —
−40 —

Write the °F temperatures.

1.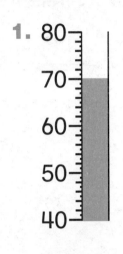

2.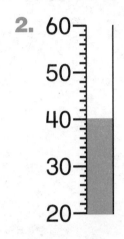

3.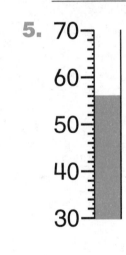

4.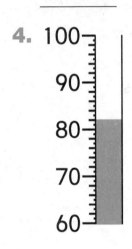

5.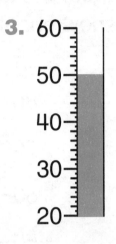

6.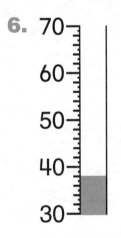

Color to show each temperature.

7. 80°F

90 —
80 —
70 —
60 —
50 —

8. 62°F

90 —
80 —
70 —
60 —
50 —

9. 58°F

90 —
80 —
70 —
60 —
50 —

Date _____

1. What is the temperature today?

_____ °F

Is the temperature odd or even?

_____

2. Write the missing numbers.

Rule

−4

| 24 | 20 | | 12 | |

3. Draw and solve.

Olga had 6 pennies.

Tyson gave her 2 more pennies.

How many pennies does Olga have now?

_____ pennies

4. Circle the winning domino in *Domino Top-It.*

Date _____

My Body and Units of Measure

Measure some objects. Record your measurements.

Unit	Picture	Object	Measurements
digit			about ____ digits
yard			about ____ yards
hand			about ____ hands
pace			about ____ paces
cubit			about ____ cubits
arm span (or fathom)			about ____ arm spans
foot			about ____ feet
hand span			about ____ hand spans

LESSON 4·2 My Height

Things that are taller than I am

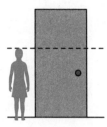

Things that are about the same size as I am

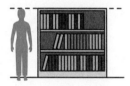

Things that are shorter than I am

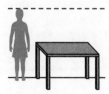

Subtracting on a Number Grid

-9	-8	-7	-6	-5	-4	-3	-2	-1	0
1	2	3	4	5	6	7	8	9	10
11	12	13	14	15	16	17	18	19	20
21	22	23	24	25	26	27	28	29	30
31	32	33	34	35	36	37	38	39	40
41	42	43	44	45	46	47	48	49	50
51	52	53	54	55	56	57	58	59	60
61	62	63	64	65	66	67	68	69	70

1. Start at 38. Count back 4. Where do you end up? _____

   38 − 4 = _____

2. Start at 59. Count back 9. Where do you end up? _____

   59 − 9 = _____

3. Start at 62. Count back 11. Where do you end up? _____

   62 − 11 = _____

4. Start at 70. Count back 17. Where do you end up? _____

   70 − 17 = _____

Try This

Subtract.

5. 43 − 20 = _____   6. 35 − 15 = _____

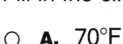 **Math Boxes**

1. What is the temperature?
Fill in the circle next to the best answer.

○ **A.** 70°F

○ **B.** 72°F

○ **C.** 80°F

○ **D.** 76°F

2. Draw the missing shape.

◇ ☐ ◇ ___ ◇ ☐

3. Complete the table.

Before	Number	After
24	25	26
	29	
	33	
	37	
	40	

4. Use a number grid.
Count by 10s.

8,   18,   ___,   ___,

___,   ___,   ___,   ___,

___,   ___,   ___

My Foot and the Standard Foot

Measure two objects with the cutout of your foot.
Draw pictures of the objects or write their names.

1. I measured

It is about _____ _____ feet.
(your name)

2. I measured

It is about _____ _____ feet.
(your name)

Measure the same two objects with the foot-long foot.
Sometimes it is called the *standard foot.*

3. I measured

It is about _____ feet.

4. I measured

It is about _____ feet.

LESSON 4·3 | **Math Boxes**

---

1. What is the temperature today?

_____ °F

Is the temperature odd or even?

_____

---

2. What comes next?

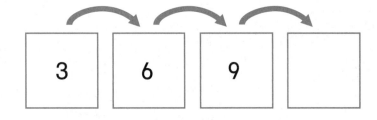

Rule
Count by 3s

| 3 | 6 | 9 | |

Fill in the circle next to the best answer.

○ **A.** 10   ○ **B.** 11   ○ **C.** 12   ○ **D.** 6

---

3. Draw and solve.

Ava had 9 pennies.

She lost 4 pennies.

How many pennies does Ava have now?

_____ pennies

---

4. Circle the winning domino in *Domino Top-It.*

LESSON 4·4   **Inches**

Pick 4 short objects to measure. Draw or name them.
Then measure the objects with your ruler.

1.

About _____ inches long

2.

About _____ inches long

3.

About _____ inches long

4.

About _____ inches long

Date _____

1. Record the temperatures.

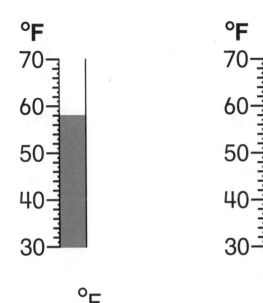

_____°F   _____°F

2. Draw the missing shape.

3. Complete the table.

Before	Number	After
27	28	29
	35	
	50	
	101	

4. Use a number grid.
Count by 10s.

/ ,   / / , _____, _____,

_____, _____, _____, _____,

_____, _____, _____

LESSON 4·5   Measuring in Inches

1. Choose two objects to measure. Estimate each object's length. Measure the objects to the nearest inch.

Object (Name it or draw it.)	My Estimate	My Measurement
	about _____ inches	about _____ inches
	about _____ inches	about _____ inches

Measure each line segment.

2. _____

about _____ inches

3. _____

about _____ inches

4. _____

about _____ inches

5. _____

about _____ inches

Draw a line segment about

6. 4 inches long.

7. 2 inches long.

Date _____

1. Measure your calculator.

It is about _____ inches long.

2. Favorite Pets, Mr. Lee's Class

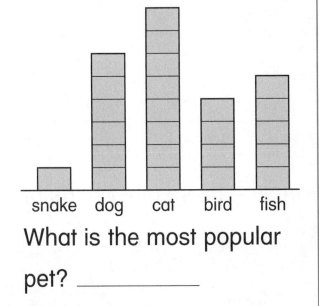

snake dog cat bird fish

What is the most popular

pet? _____

How many children like

snakes? _____

3. Count back by 5s.

40, 35, 30, _____,

_____, _____, _____,

_____

4. Write the sums.

[die] + [die] = _____

[die] + [die] = _____

[die] + [die] = _____

Date _____

Record your wrist size below.

1. Wrist

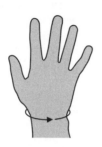

It is about _____ inches.

Measure these other parts of your body. Work with a partner.

2. Elbow

It is about _____ inches.

3. Ankle

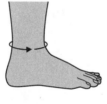

It is about _____ inches.

4. Head

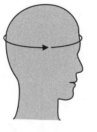

It is about _____ inches.

5. Hand span

It is about _____ inches.

LESSON 4·6

Domino Parts and Totals

Find the totals.

Example:

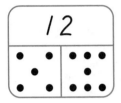

12

1.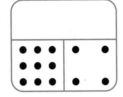

2.

3.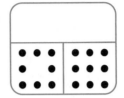

4.

Total	
Part	Part
7	4

5.

Total	
Part	Part
2	9

6.

Total	
Part	Part
6	3

7.

Total	
Part	Part
8	8

8.

Total	
Part	Part
7	7

9.

Total	
Part	Part
8	3

Try This

Find the missing parts.

10.

Total	
12	
Part	Part
6	

11.

Total	
14	
Part	Part
	5

12.

Total	
15	
Part	Part
	7

LESSON 4·6   **Math Boxes**

---

1. Draw a line segment about 3 inches long.

2. Which coins show 26¢?

   Fill in the circle next to the best answer.

   ○ **A.** Ⓝ Ⓝ Ⓟ Ⓟ Ⓟ Ⓟ Ⓟ Ⓟ

   ○ **B.** Ⓓ Ⓓ Ⓟ

   ○ **C.** Ⓓ Ⓝ Ⓝ Ⓟ

   ○ **D.** Ⓓ Ⓓ Ⓝ Ⓟ

---

3. Use your number line.

   Start at 4.

   Count up 5 hops.

   You end at _____.

   4 + 5 = _____

4. Tom has Ⓝ Ⓟ Ⓟ Ⓟ.

   Bill has Ⓓ.

   Who has more money?

   _____

   How much more money?

   _____ ¢

Date _____

1. Today's date is _____.

   My height is _____ inches.

2. This is a bar graph. It shows the heights of children in my class.

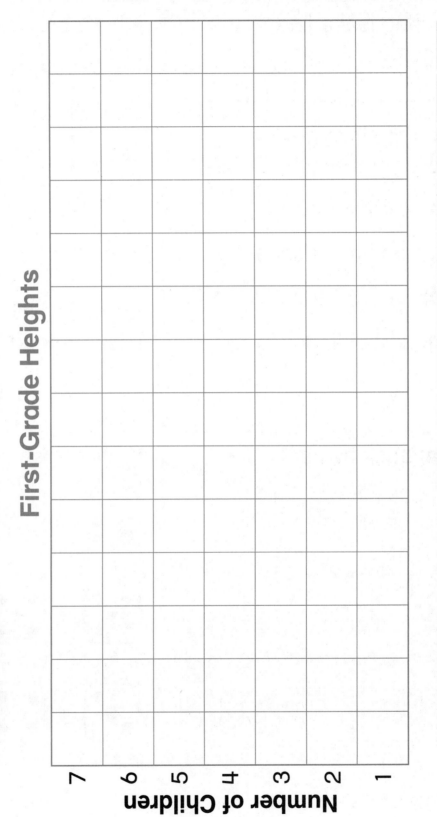

First-Grade Heights

Number of Children: 7, 6, 5, 4, 3, 2, 1

Inches Tall

3. The "typical" height for first graders in my class is about _____ inches.

Date _____

1. How long is the line segment?

_____

Fill in the circle next to the best answer.

- ○ **A.** about 3 inches
- ○ **B.** about 2 inches
- ○ **C.** about 1 inch
- ○ **D.** about 4 inches

2. Favorite Drinks, Ms. Brown's Class

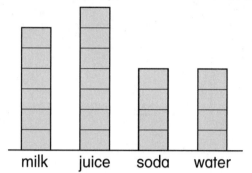

How many children like milk?

_____ children

Do more children like juice or soda?

_____

3. Count back by 2s.

36,   34,   32,

_____, _____, _____,

_____, _____, _____,

_____, _____, _____

4. Write the sums.

$\boxed{::} + \boxed{\cdot} = \underline{\hspace{1cm}}$

$\boxed{::} + \boxed{::} = \underline{\hspace{1cm}}$

$\boxed{\therefore} + \boxed{::} = \underline{\hspace{1cm}}$

LESSON 4·8 | **Telling Time**

Record the time.

1.

_____ o'clock

2.

half-past _____ o'clock

3.

quarter-past _____ o'clock

4.

quarter-to _____ o'clock

Try This

Draw the hands to show the time.

5.

half-past 3 o'clock

6.

quarter-to 5 o'clock

LESSON 4·8 **Math Boxes**

1. Draw a line segment about 2 inches long.

2. Show 47¢.

Use Ⓓ, Ⓝ, and Ⓟ.

3. Use your number line.

Start at 6.

Count up 5 hops.

You end at _____.

6 + 5 = _____

4. Nico has Ⓓ Ⓓ Ⓟ.

Kenisha has Ⓓ Ⓝ Ⓝ.

Who has more money?

_____

How much more money?

_____ ¢

School-Year Timeline

Think about these times of the year.
Draw pictures of things that happen during each of these months.

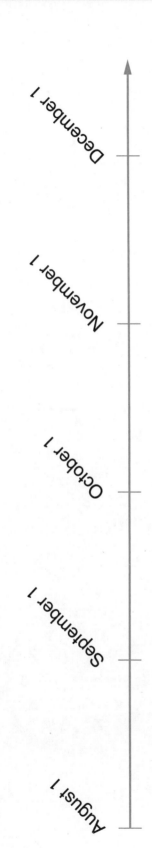

LESSON 4·9
Telling Time to the Quarter-Hour

Record the time.

1.

_____ o'clock

2.

half-past _____ o'clock

3.

quarter-after _____ o'clock

4.

quarter-after _____ o'clock

5.

quarter-to _____ o'clock

6.

quarter-to _____ o'clock

LESSON 4·9 Math Boxes

1. Measure your journal.

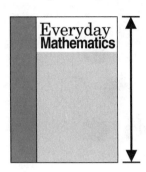

It is about _____ inches long.

2. What time is it?

Fill in the circle next to the best answer.

○ **A.** quarter-to 5 o'clock

○ **B.** quarter-to 4 o'clock

○ **C.** quarter-to 6 o'clock

○ **D.** quarter-to 9 o'clock

3. Use your number line.

Start at 8.

Count back 5 hops.

You end at _____.

$8 - 5 =$ _____

4. Write the sums.

$5 + 4 =$ _____

$6 + 3 =$ _____

LESSON 4·10 Math Boxes

1. Are you more likely to grab black or white?

_____

2. Write the missing numbers.

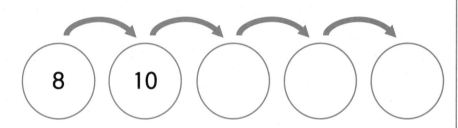

Rule

Count by 2s

8   10   ◯   ◯   ◯

3. Record the time.

quarter-after _____ o'clock

4. Make sums of 9 pennies.

Left Hand	Right Hand
4	5
7	
	1

Date _____

Draw a line from each domino to its matching numbers.
Then use the dominoes to find the sums.

1. 4 + 2 = _____

2. 3 + 1 = _____

3. _____ = 1 + 5

4. 2 + 2 = _____

5. _____ = 3 + 3

6. _____ = 1 + 4

Try This

7. Draw dots to complete the domino. Write a fact to go with it.

_____ + _____ = _____

LESSON 4·11 Math Boxes

1. Measure the line segment.

_____

It is about _____ inches long.

2. What time is it?

quarter-to _____ o'clock

3. Use your number line.

Start at 9.

Count back 7 hops.

You end at _____.

$9 - 7 =$ _____

4. Write the sums.

$1 + 5 =$ _____ $6 + 6 =$ _____

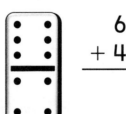 6
+ 4

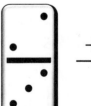 2
+ 3

LESSON 4·12 Math Boxes

1. Are you more likely to grab black or white?

_____

2. What is the rule?

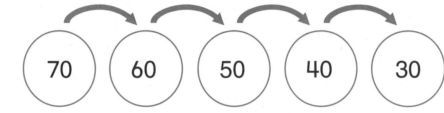

Rule
?

70 60 50 40 30

Fill in the circle next to the best answer.

○ **A.** Count up by 2s   ○ **B.** + 10

○ **C.** Subtract 5   ○ **D.** −10

3. Draw the hands.

quarter-to 6 o'clock

4. Make sums of 15 pennies.

Left Hand	Right Hand
10	5
8	
	6

LESSON
4·13   **Math Boxes**

---

1. Circle the winning domino in *Domino Top-It.*

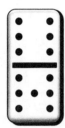

2. Tim has 10¢.

Jan has 5¢.

Who has more money?

_____

How much more money?

_____ ¢

---

3. Write the sums.

 + 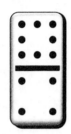 = _____

 + [image] = _____

 + [image] = _____

4. Make sums of 8 pennies.

Left Hand	Right Hand
3	5
4	
	8

Tens-and-Ones Mat

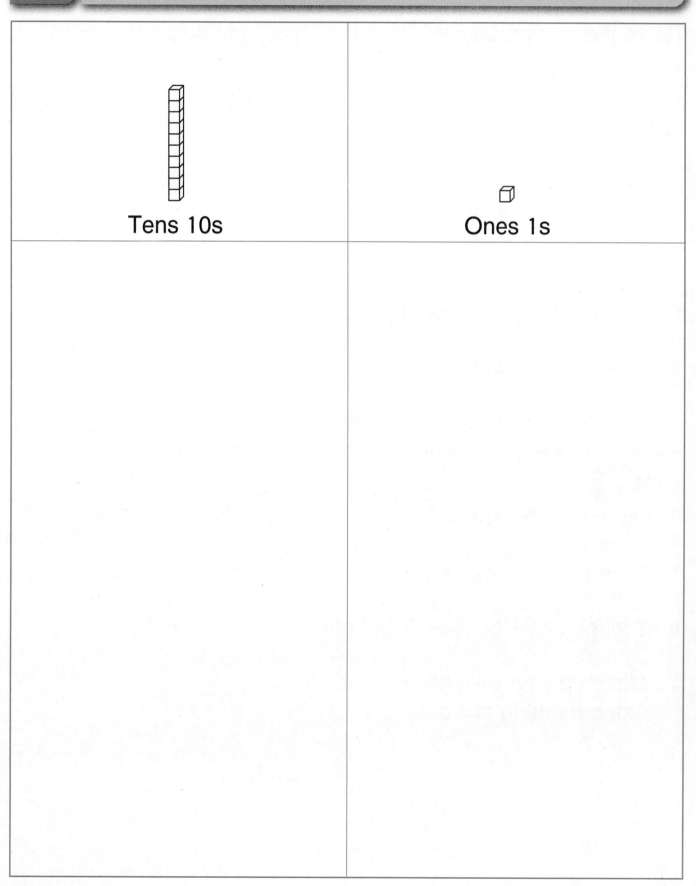

Tens 10s

Ones 1s

LESSON 5·1 Tens-and-Ones Riddles

Solve the riddles. Use your base-10 blocks to help you.

Example: 2 ▯ and 3 ◻   What am I? _23_

1. 6 ▯ and 5 ◻   What am I? _____

2. 7 ▯ and 2 ◻   What am I? _____

3. 6 longs and 4 cubes. What am I? _____

4. 7 longs and 0 cubes. What am I? _____

Try This

Trade to find the answers.

5. 1 long and 11 cubes. What am I? _____

6. 2 longs and 14 cubes. What am I? _____

7. Make up your own riddle.
   Ask a friend to solve it.

Date _____

1. Solve the riddles.

What am I? _____   What am I? _____

2. Fill in the rule and the missing numbers.

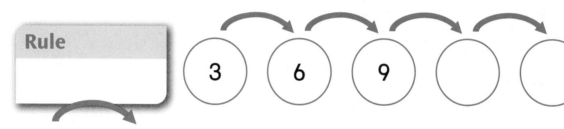

Rule _____

3   6   9   ◯   ◯

3. Find the sums.

5 + 4 = _____

_____ = 3 + 7

4. How many tallies?

||||| ||||| ||||| |||

_____ tallies

Odd or even?

_____

LESSON 5·2 Place-Value Mat

Hundreds

Tens

Ones

LESSON 5·2   **Math Boxes**

1. Solve the riddles.

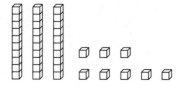

What am I? _____

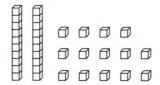

What am I? _____

2. Make a tally for 14.

_____

3. Use your number grid.

Start at 45.

Count up 13.

You end at _____.

45 + 13 = _____

4. Draw and solve.

Trey has 5 cats and 2 dogs.

How many pets does Trey have?

_____ pets

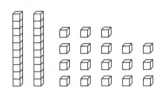

Math Boxes

1. What am I?

Choose the best answer.

⬭ 20   ⬭ 30

⬭ 38   ⬭ 218

2. Fill in the rule and the missing numbers.

Rule

2   12   22   ( )   ( )

3. Find the sums.

_____ = 2 + 5

5 + 7 = _____

4. Write the number.

HHT HHT HHT HHT HHT HHT

HHT HHT //

_____

Odd or even?

_____

LESSON 5·4 Math Boxes

1. Use | and • to show the number 42.

2. How many tallies?

‖‖‖ ‖‖‖ ‖‖‖ ‖‖‖ |||

Choose the best answer.

- ⬭ 19
- ⬭ 23
- ⬭ 25
- ⬭ 43

3. Use your number grid.

Start at 48.

Count up 15.

You end at _____.

48 + 15 = _____

4. Draw and solve.

Rosa had 9 grapes.

She ate 3 grapes.

How many grapes does Rosa have left?

_____ grapes

LESSON 5·5

Math Boxes

1. Circle the tens place.

73   52   88   15   30

2. Write the number model.

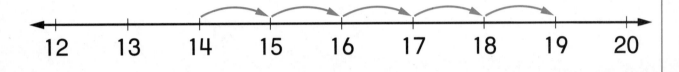

12   13   14   15   16   17   18   19   20

_____ + _____ = _____

3. Measure to the nearest inch.

_____

It is about _____ inches long.

_____

It is about _____ inches long.

4. Draw the missing dots.

Find the total number of dots.

6 + 4 = _____

7 + 3 = _____

Date _____

"Less Than" and "More Than" Number Models

Write < for "is less than" and > for "is more than."

1.  19 lb ◯ 23 lb

2. 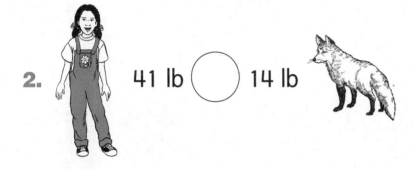 41 lb ◯ 14 lb

3. 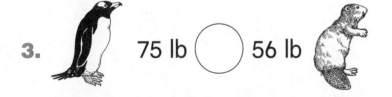 75 lb ◯ 56 lb

4. 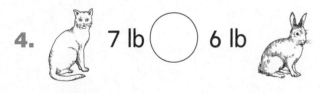 7 lb ◯ 6 lb

5. 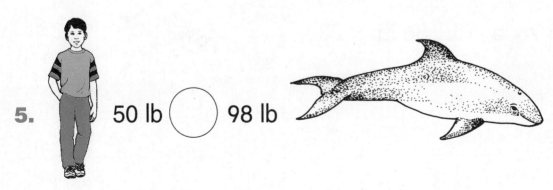 50 lb ◯ 98 lb

LESSON 5·6 "Less Than" and "More Than" Number Models *(cont.)*

Try This

Write < for "is less than" and > for "is more than."

6. 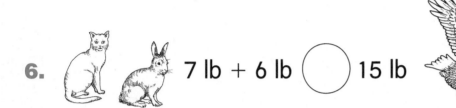 7 lb + 6 lb ◯ 15 lb

7.  120 lb ◯ 50 lb + 41 lb

8. 14 lb + 15 lb ◯ 23 lb

9. 75 lb ◯ 56 lb + 23 lb

10.  14 lb + 6 lb ◯ 19 lb

Date _____

1. Fill in the pattern.

⬜ ◎ ⬤ _____ ◎ ⬤ ⬜ ◎ _____

2. Show 32¢ with the fewest coins.

Use Ⓓ, Ⓝ, and Ⓟ.

3. **Judy's Dice Rolls**

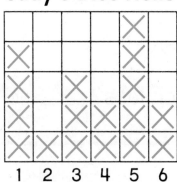

1 2 3 4 5 6
Number of Dots

How many times did Judy roll a 1?

_____ times

What number did Judy roll the most?

_____

4. Draw and solve.

The garden has 4 ladybugs and 9 ants.

How many insects are there in all?

_____ insects

LESSON 5·7 **How Much More? How Much Less?**

Find each difference.

1. John Ⓟ Ⓟ Ⓟ Ⓟ Ⓟ Ⓟ Ⓟ Ⓟ

Nick Ⓟ Ⓟ

Who has more? _____ How much more? _____¢

2. June Ⓟ Ⓟ Ⓟ Ⓟ Ⓟ Ⓟ Ⓟ Ⓟ Ⓟ

Mia Ⓟ Ⓟ Ⓟ Ⓟ Ⓟ Ⓟ

Who has less? _____ How much less? _____¢

3. Dante Ⓟ Ⓟ Ⓟ Ⓟ Ⓟ Ⓟ Ⓟ

Kala Ⓟ Ⓟ Ⓟ Ⓟ Ⓟ Ⓟ Ⓟ Ⓟ Ⓟ Ⓟ Ⓟ Ⓟ Ⓟ Ⓟ

Who has less? _____ How much less? _____¢

Try This

4. Carlos has 12 pennies.

Mary has 20 pennies.

Who has more? _____

How much more? _____¢

Date _____

---

1. Circle the ones place.

12   40   6   77   54

---

2. Write the number model.

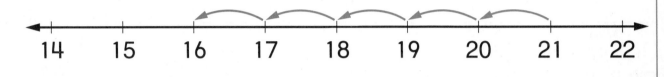

14   15   16   17   18   19   20   21   22

_____ − _____ = _____

---

3. How long is the line segment?

_____

Choose the best answer.

⬭ about 2 inches

⬭ about 3 inches

⬭ about 4 inches

⬭ about 5 inches

---

4. Draw the missing dots.

Find the total number of dots.

8 + 6 = _____

4 + 9 = _____

LESSON 5·8 Number Stories

Here is a number story Mandy made up.

I have 4 balloons.
Jamal brought 1 more.
We have 5 balloons together.

$4 + 1 = 5$

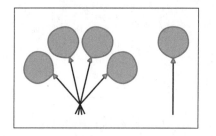

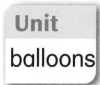

Unit
balloons

Record your own number story.
Fill in the unit box.
Write a number model.
You may want to draw a
picture for your story.

Unit

Date _____

---

1. Fill in the pattern.

   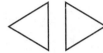

---

2. How much money?

Choose the best answer.

  ⬭ 71¢   ⬭ 61¢

  ⬭ 46¢   ⬭ 40¢

---

3.   **Teeth Lost**

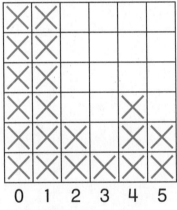

   0 1 2 3 4 5
Number of Teeth

How many children have
lost 0 teeth?

_____ children

How many children have
lost more than 3 teeth?

_____ children

---

4. Draw and solve.

The chicken has 6 eggs.

2 eggs hatch.

How many eggs are left?

_____ eggs

LESSON 5·9 Dice-Throw Record

Roll a pair of dice. Draw an X in a box for the sum, from the bottom up. Which number reached the top first? _____

2	3	4	5	6	7	8	9	10	11	12

Math Boxes

1. Write <, >, or =.

3 ☐ 13

17 ☐ 15

24 ☐ 42

28 ☐ 26

2. Draw and solve.

Meg has 8 pennies.

Maya has 3 pennies.

Who has more pennies?

_____

How many more pennies?

_____ pennies

3. How much money?

Ⓓ Ⓝ Ⓝ Ⓝ Ⓝ Ⓝ Ⓟ Ⓟ Ⓟ

_____¢

Show this amount with fewer coins.

Use Ⓟ, Ⓝ, and Ⓓ.

4. Draw the hands.

quarter-after 9 o'clock

LESSON 5·10 Turn-Around Facts Record

1 + 6 = ___	2 + 6 = ___	3 + 6 = ___	4 + 6 = ___	5 + 6 = ___	6 + 6 = ___
1 + 5 = ___	2 + 5 = ___	3 + 5 = ___	4 + 5 = ___	5 + 5 = ___	6 + 5 = ___
1 + 4 = ___	2 + 4 = ___	3 + 4 = ___	4 + 4 = ___	5 + 4 = ___	6 + 4 = ___
1 + 3 = ___	2 + 3 = ___	3 + 3 = ___	4 + 3 = ___	5 + 3 = ___	6 + 3 = ___
1 + 2 = ___	2 + 2 = ___	3 + 2 = ___	4 + 2 = ___	5 + 2 = ___	6 + 2 = ___
1 + 1 = ___	2 + 1 = ___	3 + 1 = ___	4 + 1 = ___	5 + 1 = ___	6 + 1 = ___

Math Boxes

1. Draw and solve.

Jade has 5 pennies.

Max has 9 pennies.

Who has fewer pennies?

_____

How many fewer pennies?

_____ fewer pennies

2. Write the sums.

[six dots die] + [five dots die] = _____

_____ = [four dots die] + [two dots die]

_____ = [one dot die] + [three dots die]

3. Record the temperature.

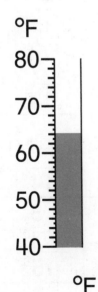

°F

_____ °F

4. Count up by 5s.

25, _____, _____,

_____, _____, _____,

_____, _____

Date

LESSON 5·11 Facts Table

0 +0 **0**	0 +1 **1**	0 +2 **2**	0 +3 **3**	0 +4 **4**	0 +5 **5**	0 +6 **6**	0 +7 **7**	0 +8 **8**	0 +9 **9**
1 +0 **1**	1 +1 **2**	1 +2 **3**	1 +3 **4**	1 +4 **5**	1 +5 **6**	1 +6 **7**	1 +7 **8**	1 +8 **9**	1 +9 **10**
2 +0 **2**	2 +1 **3**	2 +2 **4**	2 +3 **5**	2 +4 **6**	2 +5 **7**	2 +6 **8**	2 +7 **9**	2 +8 **10**	2 +9 **11**
3 +0 **3**	3 +1 **4**	3 +2 **5**	3 +3 **6**	3 +4 **7**	3 +5 **8**	3 +6 **9**	3 +7 **10**	3 +8 **11**	3 +9 **12**
4 +0 **4**	4 +1 **5**	4 +2 **6**	4 +3 **7**	4 +4 **8**	4 +5 **9**	4 +6 **10**	4 +7 **11**	4 +8 **12**	4 +9 **13**
5 +0 **5**	5 +1 **6**	5 +2 **7**	5 +3 **8**	5 +4 **9**	5 +5 **10**	5 +6 **11**	5 +7 **12**	5 +8 **13**	5 +9 **14**
6 +0 **6**	6 +1 **7**	6 +2 **8**	6 +3 **9**	6 +4 **10**	6 +5 **11**	6 +6 **12**	6 +7 **13**	6 +8 **14**	6 +9 **15**
7 +0 **7**	7 +1 **8**	7 +2 **9**	7 +3 **10**	7 +4 **11**	7 +5 **12**	7 +6 **13**	7 +7 **14**	7 +8 **15**	7 +9 **16**
8 +0 **8**	8 +1 **9**	8 +2 **10**	8 +3 **11**	8 +4 **12**	8 +5 **13**	8 +6 **14**	8 +7 **15**	8 +8 **16**	8 +9 **17**
9 +0 **9**	9 +1 **10**	9 +2 **11**	9 +3 **12**	9 +4 **13**	9 +5 **14**	9 +6 **15**	9 +7 **16**	9 +8 **17**	9 +9 **18**

LESSON 5·11   **Easy Addition Facts**

Complete.

Doubles Facts	
0 + 0 □	6 + 6 □
1 + 1 □	7 + 7 □
2 + 2 □	8 + 8 □
3 + 3 □	9 + 9 □
4 + 4 □	10 + 10 □
5 + 5 □	

10 Sums	
0 □ + ___ 10	□ + 4 10
□ + 9 10	7 □ + ___ 10
□ + 8 10	□ + 2 10
3 □ + ___ 10	□ + 1 10
□ + 6 10	10 □ + ___ 10
□ + 5 10	

LESSON 5·11 Math Boxes

1. Write <, >, or =.

6 ☐ 8

21 ☐ 12

5 + 5 ☐ 10

16 ☐ 4 + 6

2. Tina has Ⓝ Ⓝ Ⓟ Ⓟ Ⓟ Ⓟ.

Fred has Ⓝ Ⓝ Ⓝ Ⓟ Ⓟ Ⓟ.

Who has more money?

_____

How much more money?

_____¢

3. How much money?

Ⓝ Ⓓ Ⓝ Ⓟ Ⓓ Ⓝ Ⓝ

_____¢

Show this amount with fewer coins. Use Ⓟ, Ⓝ, and Ⓓ.

4. Draw the hands.

quarter-past 8 o'clock

LESSON 5·12 — "What's My Rule?"

Find the rules and missing numbers.

1.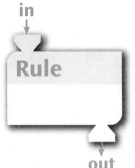

in		out
7	→	4
11	→	8
4	→	1
9	→	

Your turn: _____ → _____

2.

in		out
5	→	10
8	→	13
12	→	17
16	→	

Your turn: _____ → _____

3.

in		out
33	→	43
12	→	22
27	→	37
9	→	
24	→	

Your turn: _____ → _____

Try This

4.

in		out
1	→	2
2	→	4
3	→	6
4	→	
6	→	

Your turn: _____ → _____

LESSON 5·12 Math Boxes

1. Ray has Ⓓ Ⓝ Ⓝ Ⓟ Ⓟ.
   Dee has Ⓓ Ⓓ Ⓟ Ⓟ Ⓟ Ⓟ.

   Who has more money?

   _____

   How much more money?

   _____ ¢

2. Write the sums.

   _____ = [die 5] + [die 5]

   [die 4] + [die 6] = _____

   _____ = [die 6] + [die 3]

3. What is the temperature?

   Choose the best answer.

   ⊂⊃ 60°F   ⊂⊃ 72°F

   ⊂⊃ 68°F   ⊂⊃ 74°F

4. Count back by 5s.

   45, _____, _____,

   _____, _____, _____,

   _____, _____

LESSON 5·13 **"What's My Rule?"**

Find the rule.

1.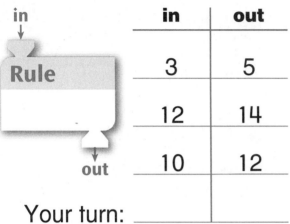

in	out
3	5
12	14
10	12

Your turn: _____

2.

in	out
4	1
12	9
17	14

Your turn: _____

3. What comes out?

Rule +10

in	out
3	13
16	
25	

Your turn: _____

4. Make your own.

Rule

in	out

Date _____

1. Write <, >, or =.

28 ☐ 38

34 ☐ 43

6 + 7 ☐ 12

16 ☐ 10 + 6

2. Lois has Ⓟ Ⓟ Ⓓ Ⓝ Ⓓ Ⓝ.
Joe has Ⓓ Ⓟ Ⓟ Ⓟ Ⓟ Ⓝ.

Who has more money?

_____

How much more money?

_____ ¢

3. How much money?

Ⓓ Ⓝ Ⓓ Ⓟ Ⓟ Ⓟ Ⓟ Ⓟ Ⓝ

_____ ¢

Show this amount with
fewer coins. Use Ⓟ, Ⓝ,
and Ⓓ.

4. It is quarter-to _____.

Choose the best answer.

⬭ 9 o'clock   ⬭ 1 o'clock

⬭ 10 o'clock   ⬭ 12 o'clock

LESSON 5·14 Math Boxes

1. Draw and solve.

Yuko has 3 red balloons, 4 green balloons, and 1 blue balloon.

How many balloons does she have in all?

_____ balloons

2. Draw the hands.

quarter-after 7 o'clock

3. Draw the missing dots.

Find the total number of dots.

$5 + 7 =$ _____

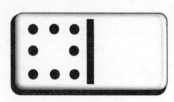

$8 + 3 =$ _____

4. Count up by 5s.

5,   10,   _____,

_____, _____, _____,

_____, _____, _____

Date _____

Notes

Date

Notes

Notes

Date _____

Date

Notes

Number Cards 0–15

15	14	13	12
11	10	9	8
7	6	5	4
3	2	1	0

Number Cards

Number Cards 16–22

16	**17**	**18**	**19**
20	**21**	**22**	**+**
−	**×**	**÷**	**=**
<	**?**	**wild card**	**wild card**

Number Cards 0–9

1 0

5 4 3 2

9 8 7 6

Clock Face, Hour and Minute

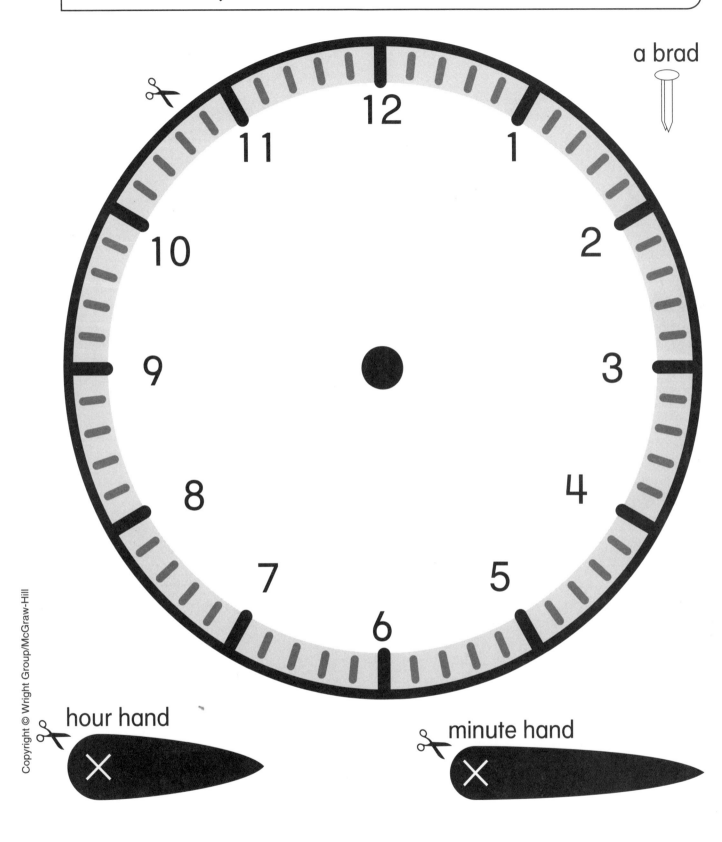

a brad

hour hand

minute hand

Activity Sheet 3

Domino Cutouts

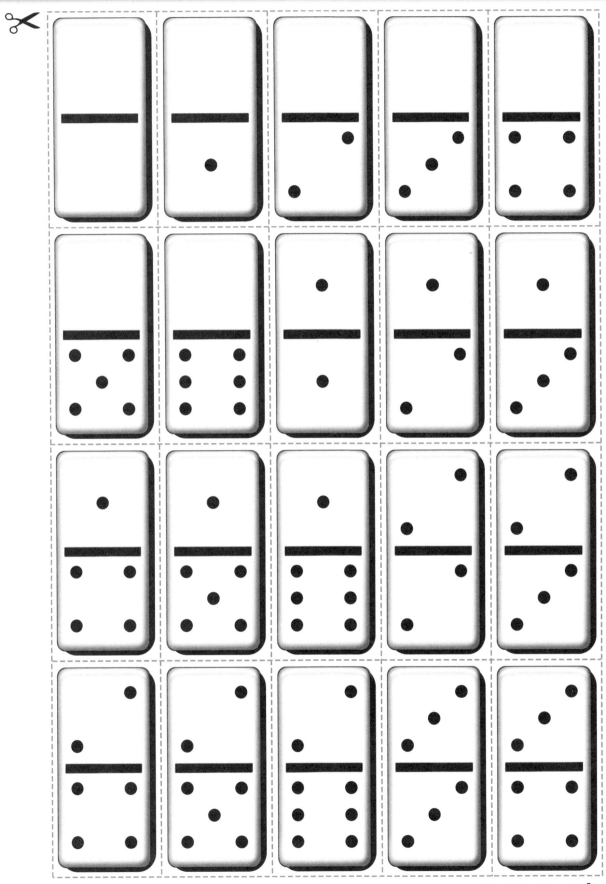

Domino Cutouts

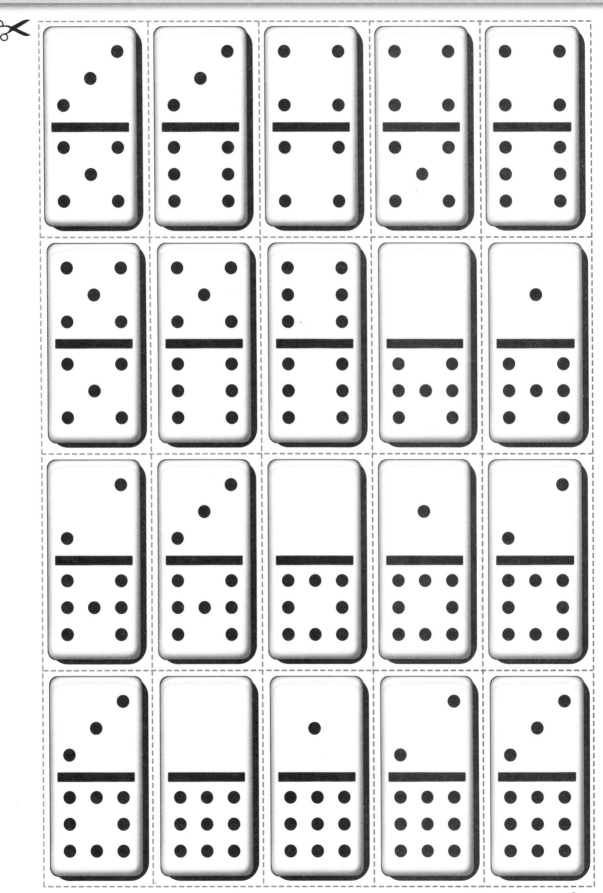

Place-Value Mat

Hundreds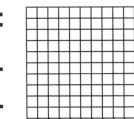

Tens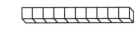

Ones

Animal Cards

First-grade girl
41 lb

7-year-old boy
50 lb

Cheetah
120 lb

Porpoise
98 lb

Penguin
75 lb

Beaver
56 lb

Animal Cards

7-year-old boy
50 in.

First-grade girl
43 in.

Porpoise
72 in.

Cheetah
48 in.

Beaver
30 in.

Penguin
36 in.

Activity Sheet 7

Animal Cards

Cat
7 lb

Fox
14 lb

Koala
19 lb

Raccoon
23 lb

Rabbit
6 lb

Eagle
15 lb

Animal Cards

Fox
20 in.

Cat
12 in.

Raccoon
23 in.

Koala
24 in.

Eagle
35 in.

Rabbit
11 in.

Activity Sheet 8